ULTIMATE SUPERCARS

FORD GT

By Ellen Labrecque

Kaleidoscope
Minneapolis, MN

The Quest for Discovery Never Ends

....

This edition first published in 2021 by Kaleidoscope Publishing, Inc.

No part of this publication may be reproduced in whole or in part without written permission of the publisher.

For information regarding permission, write to Kaleidoscope Publishing, Inc.
6012 Blue Circle Drive
Minnetonka, MN 55343

Library of Congress Control Number
2020936064

ISBN
978-1-64519-263-3 (library bound)
978-1-64519-331-9 (ebook)

Text copyright © 2021 by Kaleidoscope Publishing, Inc. All-Star Sports, Bigfoot Books, and associated logos are trademarks and/or registered trademarks of Kaleidoscope Publishing, Inc.

Printed in the United States of America.

Bigfoot lurks within one of the images in this book. It's up to you to find him!

TABLE OF CONTENTS

Chapter 1: Go, Ford, Go!... **4**

Chapter 2: Cars For Everybody.. **10**

Chapter 3: The Ford GT's Best Features....................... **16**

Chapter 4: The Ford GT Then and Now **22**

Beyond the Book... *28*

Research Ninja... *29*

Further Resources... *30*

Glossary.. *31*

Index.. *32*

Photo Credits... *32*

About the Author.. *32*

Chapter 1
Go, Ford, Go!

How cool is this? Ben's dad, Alec, tests sports cars. Alec is going to test drive the 2019 Ford GT. The Ford GT is a supercar. A supercar is a car that has a powerful engine like a race car. But it is legal to drive on public roads. Ben is joining his dad for the ride.

They are going to drive the GT on the Laguna Seca Raceway in California. The GT is liquid blue with two wide white racing stripes on the front hood of the car.

"Climb in!" Ben's dad tells him.

Ben opens the car door. It goes up in the air. "The doors look like wings!" Once Ben gets in, he can't believe how low it is to the ground. The bottom of the car is only 2.2 inches (5.6 cm) above the track!

FUN FACT
The Ford GT can reach the blistering speed of 216 miles (347 km) per hour!

PARTS OF A
FORD GT

Rear wing

Carbon fiber wheels

Alec checks Ben's seatbelt. Then he pushes a button to start the car. He presses the gas and the engine revs! Whoa! The car takes off like a spaceship. Ben is rocketed backward against his seat.

Low ride height

LED headlights

GT emblem

Dual exhaust ports support the GT's two turbo boosters. Those give the car incredible shots of horsepower.

7

Alec speeds into the first turn on the track. It feels like the tires stick to the road like glue. The sound of the engine blows Ben away. The Ford GT has a **V6** turbocharged engine. A turbocharged engine is a lot more powerful than a regular engine. Ben can feel its force through his seat.

LAGUNA SECA

Laguna Seca is a raceway in California for cars and motorcycles. It is 2.2 miles long (3.6 km). It has 11 turns. Its most famous feature is three turns in a row called "The Corkscrew." The course was built around a dried-up lake. The raceway's name means dry lagoon in Spanish.

"The engine sound is awesome!" Ben says. "I can hear it whistling!"

As they steer into the straightaway, Alec gives the car more gas. He presses his right foot on the pedal. The digital speed numbers keep climbing.

"This is awesome," Ben says. He doesn't want their ride to ever end.

Chapter 2
Cars For Everybody

Henry Ford was born on a farm near Dearborn, Michigan, in 1863. Young Henry loved to take watches and clocks apart. He liked to see how things worked. He liked to take machines apart. He also liked to build new things.

In 1896, Henry built his first automobile. This was one of the first cars ever made. It was powered by gas. It had a frame and four bicycle tires. It had a steering lever instead of a wheel. Seven years later, Henry started the Ford Motor Company. He wanted to build cars so that average people could afford to own one.

Some Model Ts are still running!

The Ford company did things differently than other car businesses. Other companies bought parts from other companies to build their cars. Henry decided to makes all the parts for his cars himself. Henry began using a system called **mass production**.

Henry Ford changed how cars are built.

WHERE THE FORD GT IS MADE

Detroit, Michigan: Ford Headquarters

FUN FACT

The letters GT stand for Gran Turismo. These are Italian words for "grand tour."

Assembly lines like this make nearly all the cars in the world.

Henry Ford probably would not recognize today's super-sleek, high-performance Ford cars like the GT.

Henry's employees worked on an **assembly line**. Each person was in charge of one part. They put their part into the car. Then the next person in line would do the same. At the end of the line, the car was complete. Henry was able to build a lot of cars faster.

His car prices were low. Regular Americans could afford to buy his cars. Before Ford, only rich people could afford cars.

Henry Ford died in 1947 at age 83. The Ford Motor Company is still going strong today.

Chapter 3
The Ford GT's Best Features

Ben loved test driving the Ford GT with his dad. But he still wants to learn more about the car. His dad decides to takes Ben to a car show. The new Ford GT is there.

Ben and Alec check out the engine. To Ben's surprise, it is in the back of the car, not the front. The engine says Ecoboost on it. This means the engine uses less gas than other cars. There are two giant holes in the back of the car. At high speeds, air passes through these tunnels. It helps the car move even faster. There is a rear wing on the back of the car. When the car is in race track mode, the wing goes up. This creates more downforce. This makes the car speedier.

FUN FACT
The GT can go from 0 to 60 miles per hour in 3 seconds!

The front of the new GT has **LED** headlights. The design on them is so cool. They look like artwork. They are also brighter than regular car lights.

Ben sits in the driver's seat. He notices the steering wheel isn't round. It is actually flat on the top and the bottom. Some race cars have this setup. The wheel also has lots of buttons. Everything a driver could need is right there at his fingertips. This car isn't like regular Ford cars. It is way cooler.

The GT's rear spoiler helps it cut through the air smoothly.

Check out those doors! Unlike in most cars, the GT's doors rise upward, not outward. Once inside, passengers push a button to bring the doors back down.

THE FORD GT IN DETAIL

Height: 3 feet, 8 inches (1.13 m)

Width: 6 feet, 7 inches (1.9 meters)

COST: $500,000 (United States)

LENGTH: 15 feet, 7 inches (4.8 m)

WEIGHT: 3,705 pounds (1,681 kg)

TOP SPEED: 216 miles per hour (348 kph)

TIME FROM 0-60 MPH: 3 seconds

The seats in the GT don't move up or back. Instead, the gas and brake pedals move. The steering wheel also moves. Ben sits back in the driver's seat. He wishes he were old enough to take this car on the open road. "Someday," his dad tells him. *Someday can't come soon enough*, Ben thinks.

Chapter 4
The Ford GT Then and Now

The Ford GT is the greatest sports car Ford ever made. The first model was built in 1964. It was called a GT40. The Ford Company wanted to make cars that could win car races. In particular, it wanted to win the famous Le Mans race. It wanted to defeat another famous car company, Ferrari. Ford and Ferrari became racing **rivals**. The Ford GT beat Ferrari for the first time in Le Mans in 1966. Before that, Ferrari had won Le Mans the previous five years in a row.

The driver of a Ford GT can see that number pop up on the speedometer. But only on a race track!

FUN FACT
The 40 in GT40 refers to the 40 inch height of the race car.

Ford began making the Ford GT again in 2004. They made this version of the GT for only three years. In 2016, they began making a second generation of Ford GT's. They chose to make the GT again in 2016 for a simple reason. It was 50 years since Ford beat Ferrari at Le Mans for the first time.

THE LE MANS

The 24 Hours of Le Mans is the world's oldest active automobile endurance race. It is held in Le Mans, France. Teams race around an 8.5-mile (13.8 km) track as many times as they can during a period of 24 hours. Teams usually use three drivers. They cover distances over 3,000 miles (5,000 km). A Ford GT has won this race four times (1966-69).

Ford plans to make a new GT every year through 2022. They hope each car will be faster than the previous version. Car fans better not blink. Otherwise, they'll miss the new Ford GT zooming by.

The Ford GT heads to the finish line in a race.

The steering wheel is not round. It's designed for racers.

BEYOND THE BOOK

After reading the book, it's time to think about what you learned. Try the following exercises to jumpstart your ideas.

RESEARCH

FIND OUT MORE. Where would you go to find out more about your favorite cars? Find out what company makes the car and locate its website. What information do the companies provide? What other sources of car information can you find?

CREATE

GET ARTISTIC. Cars start with creative artists and designers. Time for you to take a shot! Get art materials and create a great, new car. Will you make it a sports car? A sedan? A race car? What colors will you paint it? What features can you give it? Let your imagination go for a spin!

DISCOVER

DIG DEEPER. Henry Ford was a groundbreaking carmaker. He helped make cars affordable and available to millions of people. He also had some controversial opinions about workers and about society. Do some research on Ford and find out more about him.

GROW

GO TO A CAR SHOW. Car shows are a great way to see lots of cool cars up-close. Check your local events calendar, or ask at a car dealer for upcoming events. You can find shows of old cars and new cars, sports cars and classic cars. Go to a show and find a new favorite car to love!

RESEARCH NINJA

Visit *www.ninjaresearcher.com/2633* to learn how to take your research skills and book report writing to the next level!

RESEARCH

DIGITAL LITERACY TOOLS

SEARCH LIKE A PRO
Learn about how to use search engines to find useful websites.

FACT OR FAKE?
Discover how you can tell a trusted website from an untrustworthy resource.

TEXT DETECTIVE
Explore how to zero in on the information you need most.

SHOW YOUR WORK
Research responsibly—learn how to cite sources.

WRITE

GET TO THE POINT
Learn how to express your main ideas.

PLAN OF ATTACK
Learn prewriting exercises and create an outline.

DOWNLOADABLE REPORT FORMS

Further Resources

BOOKS

Barlow, Jason. *Top Gear Ultimate Supercars*. London, England: BBC Books, 2019.

Lamm, John. *Supercar Revolution: The Fastest Cars of All Time*. Minneapolis, MN: Motorbooks, 2018.

Streather, Adrian. *Ford GT: Then and Now*. Dorchester, United Kingdom: Veloce Publishing, 2017.

WEBSITES

FACTSURFER

Factsurfer.com gives you a safe, fun way to find more information.

1. Go to www.factsurfer.com.
2. Enter "Ford GT" into the search box and click
3. Select your book cover to see a list of related websites.

Glossary

assembly line: a place in which factory workers each add a part to make one thing.

LED: (Light Emitting Diode) a type of light that is brighter and more energy efficient than a regular light.

mass production: the making of machinery in big quantity.

rivals: people or teams in competition to be the best in a field.

V6: an engine that is shaped like a V with six cylinders.

Index

24 Hours of Le Mans, 22, 24, 25
doors, 5, 19
Ecoboost, 16
engine, 4, 6, 8, 9, 16
Ferrari, 22, 24
Ford, Henry, 10, 12, 14, 15
Ford Motor Company, 10, 12, 15, 22
France, 25
GT40, 22, 23
interior, 18, 19, 21
Laguna Seca Raceway, 5, 8
Michigan, 10, 13
Model T, 11

PHOTO CREDITS

The images in this book are reproduced through the courtesy of: Ford Media Center: 6, 14 inset, 18, 20, 21, 22, 23, 24, 26. Newscom: Larry Placido/Icon SW 8. Shutterstock: JK Multimedia 4; Ruben Ramos 10; Alexander Chizhenok 14; Zoran Karapancev 16.
Cover: Darren Brode/Shutterstock (car); gyn9037/Shutterstock (background, top); zhao jiankang/Shutterstock (background, bottom).

About the Author

Ellen Labrecque has written many non-fiction books for kids. Previously, she was a senior editor at *Sports Illustrated Kids* where she enjoyed following the race car scene. She lives outside of Philadelphia with her husband, two kids, and her dog Oscar — who also happens to be the best writing partner ever.